ADDITION

Written By:

Kim Mitzo Thompson

Karen Mitzo Hilderbrand

Illustrations:
Goran Kozjak
Mark Paskiet

Cover Design:
Steve Ruttner

Layout:
Jeanna Taipale

Twin 202 - **ADDITION** - ISBN # 0-9632249-2-1

TABLE OF CONTENTS

ADDITION WORKSHEETS:

BRAINBUSTERS:

PROBLEM SOLVING:

TIME TESTS:

AWARDS:

Name _____

Addition Facts: Sums 0 - 10

Solve each addition problem.

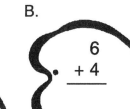

A.
```
  3
+ 2
____
```

B.
```
  6
+ 4
____
```

C.
```
  4
+ 3
____
```

D.
```
  5
+ 2
____
```

E.
```
  2
+ 8
____
```

F.
```
  4
+ 4
____
```

G.
```
  2
+ 2
____
```

H.
```
  1
+ 4
____
```

I.
```
  3
+ 6
____
```

J.
```
  9
+ 1
____
```

K.
```
  5
+ 3
____
```

L.
```
  7
+ 2
____
```

M.
```
  8
+ 2
____
```

N.
```
  3
+ 3
____
```

O.
```
  2
+ 6
____
```

P.
```
  3
+ 7
____
```

Q.
```
  4
+ 5
____
```

R.
```
  10
+ 0
____
```

S.
```
  3
+ 4
____
```

T.
```
  5
+ 5
____
```

U.
```
  2
+ 3
____
```

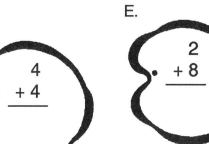

© 1991 Twin Sisters Productions, Inc.

3

TWIN 202 - Addition

Name _____

Solve each addition problem.

A. 2 + 2 =

B. 2 + 3 =

C. 2 + 6 =

D. 3 + 4 =

E. 3 + 6 =

F. 4 + 5 =

G. 4 + 2 =

H. 5 + 1 =

I. 5 + 3 =

J. 5 + 5 =

K. 6 + 2 =

L. 6 + 4 =

M. 7 + 2 =

N. 8 + 0 =

O. 9 + 1 =

TWIN 202 - Addition

Name _____

Solve each addition problem.

Addition Facts:
Sums 0 - 12

6 + 6 =

4 + 2 =

7 + 3 =

9 + 2 =

$\begin{array}{r} 8 \\ + 4 \\ \hline \end{array}$

$\begin{array}{r} 5 \\ + 3 \\ \hline \end{array}$

$\begin{array}{r} 2 \\ + 3 \\ \hline \end{array}$

$\begin{array}{r} 3 \\ + 3 \\ \hline \end{array}$

$\begin{array}{r} 7 \\ + 4 \\ \hline \end{array}$

$\begin{array}{r} 5 \\ + 4 \\ \hline \end{array}$

$\begin{array}{r} 4 \\ + 6 \\ \hline \end{array}$

$\begin{array}{r} 5 \\ + 6 \\ \hline \end{array}$

$\begin{array}{r} 7 \\ + 2 \\ \hline \end{array}$

$\begin{array}{r} 8 \\ + 2 \\ \hline \end{array}$

$\begin{array}{r} 5 \\ + 5 \\ \hline \end{array}$

$\begin{array}{r} 2 \\ + 5 \\ \hline \end{array}$

$\begin{array}{r} 2 \\ + 2 \\ \hline \end{array}$

$\begin{array}{r} 4 \\ + 4 \\ \hline \end{array}$

$\begin{array}{r} 5 \\ + 7 \\ \hline \end{array}$

$\begin{array}{r} 4 \\ + 3 \\ \hline \end{array}$

$\begin{array}{r} 3 \\ + 8 \\ \hline \end{array}$

$\begin{array}{r} 3 \\ + 7 \\ \hline \end{array}$

8 + 3 =

$\begin{array}{r} 2 \\ + 9 \\ \hline \end{array}$

7 + 5 =

$\begin{array}{r} 4 \\ + 7 \\ \hline \end{array}$

$\begin{array}{r} 3 \\ + 5 \\ \hline \end{array}$

$\begin{array}{r} 4 \\ + 5 \\ \hline \end{array}$

5

Twin 202 - Addition

Name _____

Add. Then color your answers in the picture below.

A.

3 + 4 = _____
2 + 1 = _____
7 + 4 = _____
8 + 2 = _____
1 + 1 = _____

B.

5 + 6 = _____
3 + 2 = _____
0 + 1 = _____
2 + 7 = _____
5 + 1 = _____

C.

4 + 4 = _____
7 + 5 = _____
2 + 2 = _____
6 + 6 = _____
3 + 7 = _____

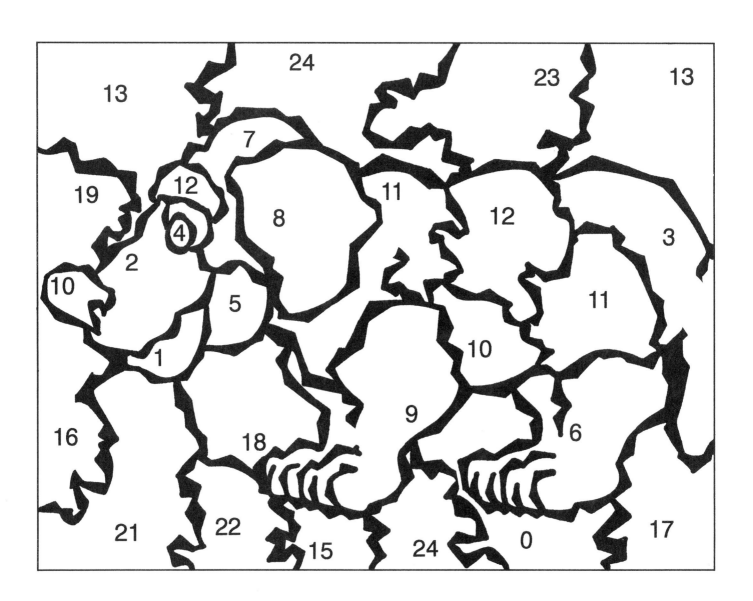

Name _____

Solve each addition problem.

A. 5 + 2 =

B. 9 + 3 =

C. 5 + 4 =

D. 6 + 5 =

E. 4 + 6 =

F. 7 + 2 =

G. 3 + 3 =

H. 2 + 2 =

I. 3 + 5 =

J. 2 + 3 =

K. 3 + 7 =

L. 8 + 3 =

M. 8 + 4 =

N. 4 + 4 =

O. 6 + 6 =

P. 7 + 3 =

Q. 2 + 6 =

R. 4 + 2 =

S. 5 + 5 =

T. 9 + 2 =

Twin 202 - Addition

Name _____

Add to solve the problem.

$4 + 6 =$

$7 + 5 =$

$6 + 1 =$

$2 + 1 =$

$3 + 3 =$

$2 + 6 =$

$2 + 9 =$

$3 + 6 =$

$3 + 5 =$

$5 + 6 =$

$4 + 4 =$

$8 + 2 =$

$6 + 6 =$

$3 + 9 =$

$8 + 3 =$

$2 + 3 =$

$4 + 3 =$

B.

A.

C.

$2 + 2 =$

8

TWIN 202 - Addition

Name _____

Add to solve the problem.

7 + 9 =

8 + 4 =   9 + 5 =

5 + 6 =

4 + 6 =

7 + 3 =

1 + 5 =

9 + 9 =

6 + 6 =

6 + 4 =

12 + 5 =

10 + 5 =

8 + 8 =

5 + 9 =

6 + 7 =

3 + 3 =

12 + 9 =

11 + 6 =

5 + 5 =

8 + 2 =

4 + 8 =

6 + 3 =   7 + 4 =

4 + 3 =

5 + 7 =

Name _____

Example:

Circle another name for:

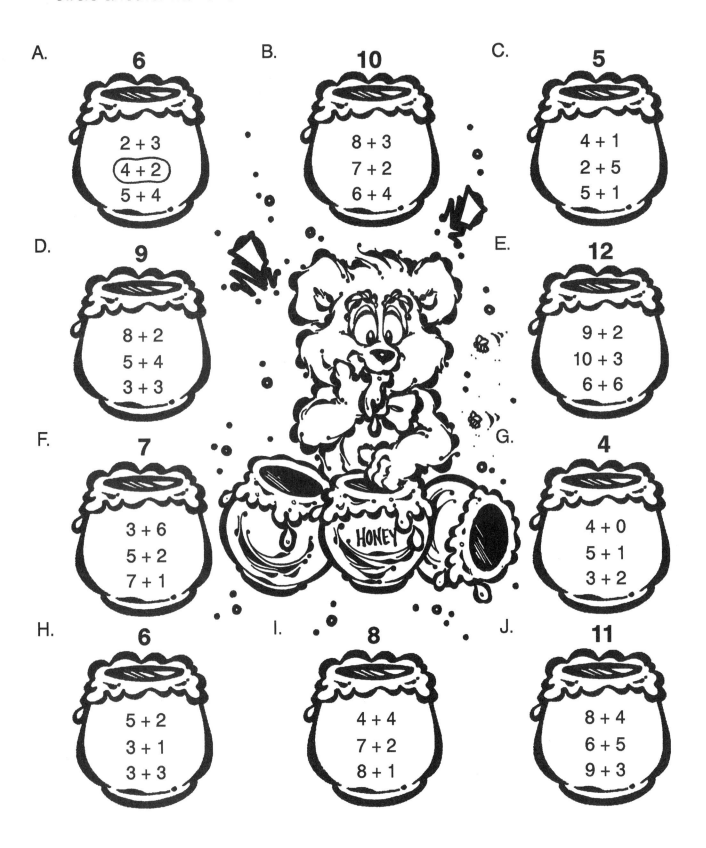

A. **6**

2 + 3
(4 + 2)
5 + 4

B. **10**

8 + 3
7 + 2
6 + 4

C. **5**

4 + 1
2 + 5
5 + 1

D. **9**

8 + 2
5 + 4
3 + 3

E. **12**

9 + 2
10 + 3
6 + 6

F. **7**

3 + 6
5 + 2
7 + 1

HONEY

G. **4**

4 + 0
5 + 1
3 + 2

H. **6**

5 + 2
3 + 1
3 + 3

I. **8**

4 + 4
7 + 2
8 + 1

J. **11**

8 + 4
6 + 5
9 + 3

TWIN 202 - Addition

Name _____

Add. Then color your answers in the picture below.

A.

6 + 6 = _____

3 + 2 = _____

2 + 1 = _____

1 + 0 = _____

8 + 6 = _____

9 + 9 = _____

B.

4 + 2 = _____

5 + 5 = _____

4 + 3 = _____

2 + 2 = _____

10 + 6 = _____

12 + 5 = _____

C.

4 + 5 = _____

7 + 4 = _____

1 + 1 = _____

6 + 2 = _____

8 + 5 = _____

9 + 4 = _____

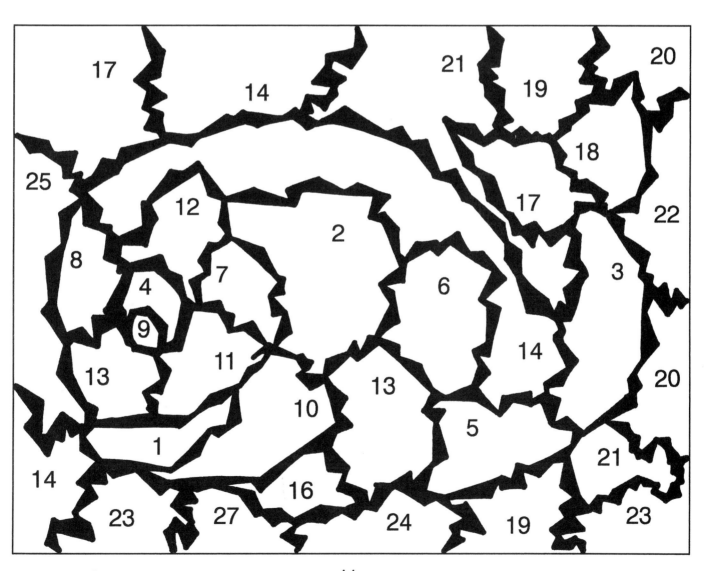

Name _____

Fill in the correct number to make each problem equal.

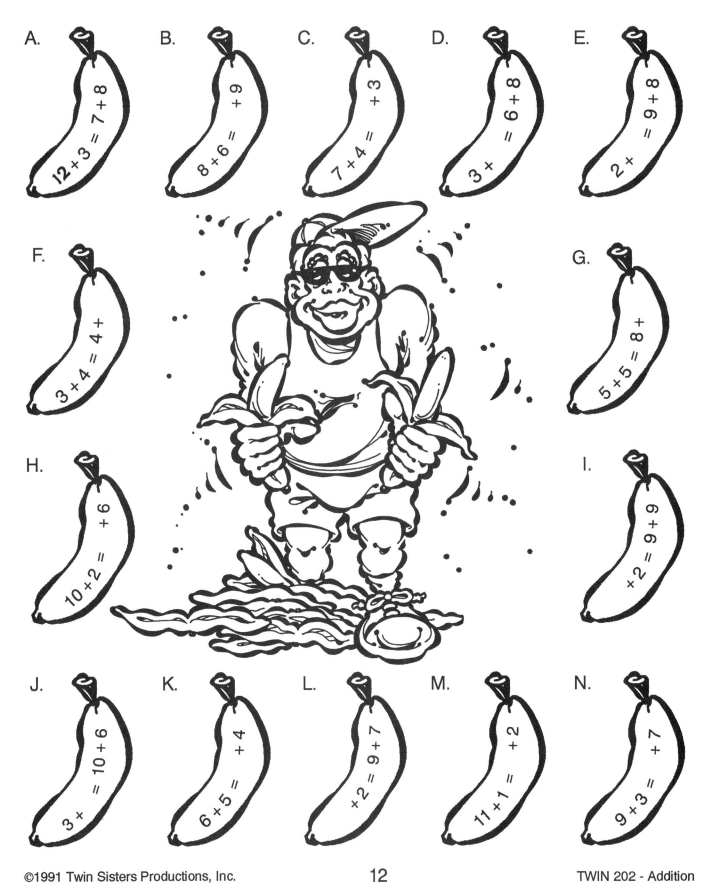

A. $12 + 3 = 7 + 8$

B. $8 + 6 = \underline{} + 9$

C. $7 + 4 = \underline{} + 3$

D. $3 + \underline{} = 6 + 8$

E. $2 + \underline{} = 9 + 8$

F. $3 + 4 = 4 + \underline{}$

G. $5 + 5 = 8 + \underline{}$

H. $10 + 2 = \underline{} + 6$

I. $\underline{} + 2 = 9 + 9$

J. $3 + \underline{} = 10 + 6$

K. $6 + 5 = \underline{} + 4$

L. $\underline{} + 2 = 9 + 7$

M. $11 + 1 = \underline{} + 2$

N. $9 + 3 = \underline{} + 7$

TWIN 202 - Addition

Name _____

Solve each addition problem.

Name _____

Solve the problems in the list below. Find each problem hidden in the puzzle. Circle each problem and write an + and an = sign in the correct place.

A.

6 + 6 = _____
9 + 7 = _____
10 + 3 = _____
2 + 9 = _____
8 + 4 = _____
7 + 7 = _____
5 + 8 = _____
3 + 7 = _____
1 + 5 = _____
6 + 8 = _____

B.

9 + 9 = _____
10 + 6 = _____
3 + 8 = _____
5 + 9 = _____
7 + 8 = _____
8 + 8 = _____
4 + 5 = _____
3 + 9 = _____
6 + 4 = _____
11 + 1 = _____

6	6	12	0	5	10	6	16	4	15	6
20	9	7	9	8	18	12	6	5	18	4
8	7	10	3	13	1	2	5	9	0	10
14	16	11	1	12	10	13	9	6	10	5
6	10	6	4	6	7	20	14	11	12	11
8	12	11	5	3	8	11	7	8	2	3
14	6	1	10	12	15	7	6	1	4	9
3	1	3	7	10	8	8	16	12	8	12

14

TWIN 202 - Addition

Name _____

Add to solve the problem.

Example:

A.
```
  3
  2
+ 6
————
 11
```
(answer in fish: 5)

B.
```
  4
  5
+ 8
————
```

C.
```
  2
  5
+ 9
————
```

D.
```
 10
  2
+ 6
————
```

E.
```
 12
  2
+ 3
————
```

F.
```
 11
  4
+ 3
————
```

G.
```
  3
  6
+ 7
————
```

H.
```
  8
  2
+ 5
————
```

I.
```
  9
  8
+ 1
————
```

J.
```
 10
  2
+ 6
————
```

K.
```
  3
  3
+ 4
————
```

L.
```
 12
  1
+ 2
————
```

Name _____

Solve. Write **<** (less than) or **>** (greater than)

A. 6 + 3 ◯ 4 + 4

B. 2 + 5 ◯ 8 + 2

C. 9 + 4 ◯ 7 + 8

D. 3 + 8 ◯ 5 + 5

E. 2 + 7 ◯ 3 + 3

F. 11 + 4 ◯ 12 + 5

G. 10 + 3 ◯ 9 + 5

H. 7 + 3 ◯ 6 + 2

I. 3 + 4 ◯ 4 + 4

J. 6 + 5 ◯ 5 + 7

K. 8 + 1 ◯ 2 + 6

L. 2 + 2 ◯ 5 + 1

M. 5 + 4 ◯ 6 + 4

N. 2 + 3 ◯ 4 + 2

O. 1 + 9 ◯ 5 + 4

P. 8 + 3 ◯ 9 + 4

Q. 7 + 3 ◯ 11 + 0

R. 5 + 5 ◯ 2 + 6

S. 2 + 2 ◯ 1 + 2

T. 10 + 4 ◯ 6 + 5

U. 8 + 8 ◯ 9 + 8

V. 7 + 6 ◯ 8 + 9

W. 9 + 0 ◯ 6 + 10

X. 12 + 3 ◯ 11 + 5

I like to eat the largest number.

example:

6 + 2 ◯ 7 + 3

TWIN 202 - Addition

Name _____

Add to solve the problem. Cross off the answer on the hippopotamus.

A.
```
  23
+ 62
-----
```

B.
```
  71
+ 20
-----
```

C.
```
  52
+ 23
-----
```

D.
```
  43
+ 31
-----
```

E.
```
  65
+ 14
-----
```

F.
```
  66
+ 23
-----
```

G.
```
  38
+ 61
-----
```

H.
```
  24
+ 44
-----
```

I.
```
  37
+ 32
-----
```

J.
```
  76
+ 10
-----
```

K.
```
  35
+ 14
-----
```

L.
```
  50
+ 40
-----
```

M.
```
  85
+ 13
-----
```

N.
```
  44
+ 22
-----
```

O.
```
  41
+ 36
-----
```

P.
```
  34
+ 53
-----
```

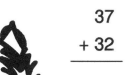

Do Not Feed Hippos

91	74	99	66
75	86	90	77
89	68	49	87
98	69	79	85

Twin 202 - Addition

Name _____

Addition of 2 Digit Numbers,
Without Regrouping

Add to solve the problem. Cross off the answer on the giraffe.

A.
38
+ 11

B.
40
+ 39

C.
62
+ 34

D.
12
+ 25

E.
81
+ 17

F.
71
+ 26

G.
22
+ 33

H.
50
+ 35

I.
24
+ 51

J.
49
+ 20

K.
10
+ 34

L.
29
+ 60

M.
34
+ 21

N.
26
+ 31

O.
30
+ 40

P.
62
+ 11

Q.
71
+ 20

R.
44
+ 44

S.
12
+ 14

T.
68
+ 10

©1991 Twin Sisters Productions, Inc.

18

TWIN 202 - Addition

Name _____

Add to solve the problems.

A.
```
  59
+ 28
-----
```

B.
```
  32
+ 48
-----
```

C.
```
  26
+ 34
-----
```

D.
```
  54
+ 27
-----
```

E.
```
  49
+ 42
-----
```

F.
```
  37
+ 24
-----
```

G.
```
  48
+ 17
-----
```

H.
```
  66
+ 27
-----
```

I.
```
  79
+ 13
-----
```

J.
```
  27
+ 57
-----
```

K.
```
  29
+ 35
-----
```

L.
```
  44
+ 26
-----
```

M.
```
  75
+ 15
-----
```

N.
```
  67
+ 19
-----
```

O.
```
  24
+ 38
-----
```

P.
```
  68
+ 19
-----
```

Q.
```
  18
+ 25
-----
```

R.
```
  46
+ 49
-----
```

S.
```
  27
+ 37
-----
```

T.
```
  68
+ 18
-----
```

U.
```
  22
+ 58
-----
```

V.
```
  37
+ 48
-----
```

W.
```
  55
+ 26
-----
```

X.
```
  42
+ 19
-----
```

Y.
```
  61
+ 29
-----
```

Twin 202 - Addition

Name _____

Add to solve the problems. Use the code to find the correct message.

CODE												
80	64	95	48	87	52	84	28	76	46	33	65	51
M	G	A	R	I	O	E	T	L	H	S	C	N

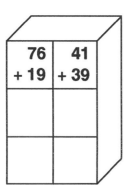

59
+ 28

76	41
+ 19	+ 39

48	65	39	29	22	38	17	26
+ 28	+ 19	+ 56	+ 19	+ 29	+ 49	+ 34	+ 38

67	19
+ 28	+ 9

16	48	28	29	36	57
+ 17	+ 17	+ 18	+ 23	+ 16	+ 19

2 + 2 = 4

TWIN 202 - Addition

Name _____

Add to solve the problems. Cross off the answers in the mud puddle.

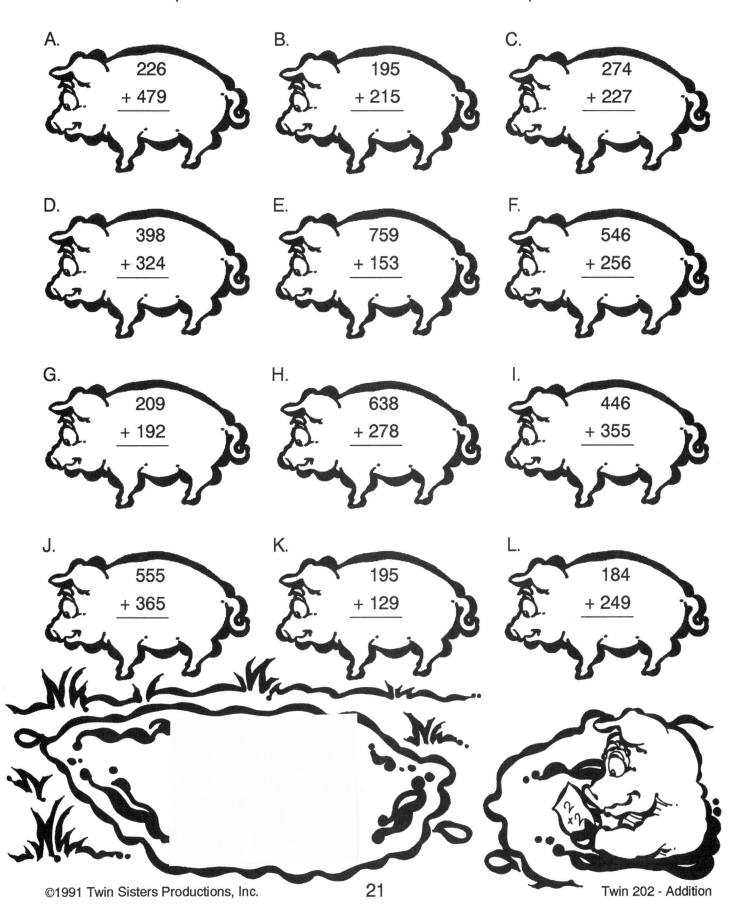

A.
226
+ 479

B.
195
+ 215

C.
274
+ 227

D.
398
+ 324

E.
759
+ 153

F.
546
+ 256

G.
209
+ 192

H.
638
+ 278

I.
446
+ 355

J.
555
+ 365

K.
195
+ 129

L.
184
+ 249

Name _____

Add to solve the problems. Cross off the answer on the lion.

A.	B.	C.	D.	E.
385 + 126	749 + 232	459 + 149	284 + 238	175 + 438

F.	G.	H.	I.	J.
253 + 167	809 + 158	599 + 232	465 + 287	394 + 129

K.	L.	M.	N.	O.
846 + 124	145 + 365	824 + 139	628 + 367	125 + 496

P.	Q.	R.	S.	T.
269 + 524	199 + 122	654 + 163	348 + 158	284 + 339

621	420	608	522	817
523	981	963	995	793
623	967	831	752	321
613	511	510	506	970

ROAR

22

TWIN 202 - Addition

Name _____

Transportation Tales

Solve each story problem.

A. **Example:**

5 trains

2 cars

How many altogether?

$$\begin{array}{r} 5 \\ +\ 2 \\ \hline 7 \end{array}$$

B.

4 airplanes

4 boats

How many altogether?

C.

3 blue tractors

2 red tractors

How many altogether?

D.

9 orange bicycles

1 green bicycle

How many altogether?

E.

4 yellow buses

2 pink buses

How many altogether?

F.

7 black horses

2 white horses

How many altogether?

G.

2 hot-air balloons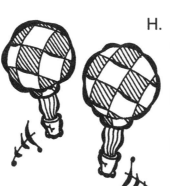

2 big blimps

How many altogether?

H.

1 black wagon

2 purple wagons

How many altogether?

     Twin 202 - Addition

Playground Fun

Solve each story problem.

A.

    6 children playing baseball   ____ + ____ = _____

    4 children join in

How many children altogether?

B.

    9 boys playing tag   ____ + ____ = _____

    3 girls jumping rope

How many children playing?

C.

    4 red sliding boards   ____ + ____ = _____

    2 yellow swings

How many playthings altogether?

D.

    8 children on teeter totters   ____ + ____ = _____

    2 children join in

How many children playing altogether?

E.

    3 teachers watching children   ____ + ____ = _____

    1 teacher joins them

How many teachers watching children?

F.

    5 boys on the monkey bars   ____ + ____ = _____

    7 girls join them

How many children playing altogether?

G.

    11 children playing soccer   ____ + ____ = _____

    6 children playing football

How many children playing altogether?

          TWIN 202 - Addition

Barnyard Madness

Solve each story problem.

A.

2 pigs eating
4 pigs sleeping
How many pigs altogether?

___ + ___ = _____

B.

6 cows grazing
5 cows join them
How many cows altogether?

___ + ___ = _____

C.

3 yellow ducks waddling
7 more join them
How many ducks altogether?

___ + ___ = _____

D.

8 brown horses
4 white horses
How many horses altogether?

___ + ___ = _____

E.

2 chickens clucking
7 chickens eating
How many chickens altogether?

___ + ___ = _____

F.

1 farmer plowing
2 farmers picking
How many farmers altogether?

___ + ___ = _____

Vacationing At The Beach

Solve each story problem.

A. Tyler built 2 sandcastles. His brother Robert built 3 sandcastles. How many sandcastles were built altogether?

___ + ___ = _____

B. Sue caught 6 minnows. Jennifer caught 5 minnows. How many minnows were caught in all?

___ + ___ = _____

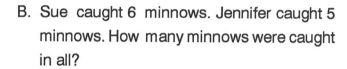

C. Seven swimmers were floating. Two swimmers joined them. How many swimmers were floating in all?

___ + ___ = _____

D. Jason had 9 yellow beachballs. Nicole had 8 blue beachballs. How many beachballs were there altogether?

___ + ___ = _____

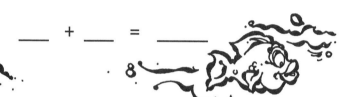

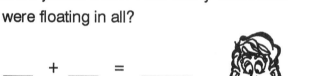

E. Four girls were diving. Six boys joined them. How many children were diving altogether?

___ + ___ = _____

F. Eleven swimmers were wearing flippers. Five more put flippers on. How many swimmers were wearing flippers in all?

___ + ___ = _____

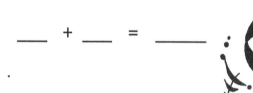

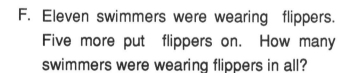

G. Judy and Kathy were snorkeling. Five more children joined them. How many children were snorkeling altogether?

___ + ___ = _____

H. Joshua caught 12 starfish. Phillip caught 6 starfish. How many starfish were caught in all?

___ + ___ = _____

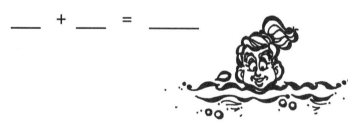

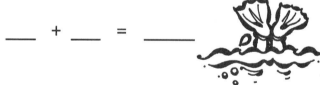

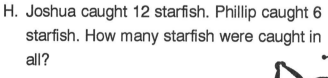

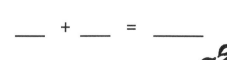

Amusement Park Adventures

Solve each story problem.

A.
Twelve children were eating cotton candy. Five adults were eating popcorn. How many people were eating altogether?

_____ = _____

B.
There are 4 roller coasters at the front of the park. There are 6 roller coasters at the back of the park. How many roller coasters are there in all?

_____ = _____

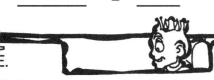

F.
Christopher went down the big slide 6 times. Anthony went down the big slide 5 times. How many times did they go down the big slide altogether?

_____ = _____

E.
Eight parents went on the ferris wheel. Ten children went with them. How many people went on the ferris wheel altogether?

_____ = _____

C.
Nine boys like chocolate ice cream. Seven girls like chocolate ice cream. How many children like chocolate ice cream in all?

_____ = _____

D.
Karen won 3 stuffed animals. Kim won 2 stuffed animals. How many stuffed animals did they win altogether?

_____ = _____

Stuffed Animals For Sale

How much for both? 3.25

A. **Example**

$3.25
+ 5.50
———
$8.75

B.

C.

D.

E.

F.

28   TWIN 202 - Addition

Name _____

WOW

_____
PROBLEMS
CORRECT

4	2	8	1	5	7	6	3	2	4
+ 2	+ 1	+ 2	+ 6	+ 5	+ 2	+ 1	+ 3	+ 5	+ 6

3	3	7	4	7	5	2	1	6	2
+ 2	+ 3	+ 3	+ 1	+ 0	+ 1	+ 2	+ 9	+ 2	+ 5

5	3	9	2	7	5	4	8	3	7
+ 2	+ 5	+ 1	+ 4	+ 1	+ 4	+ 3	+ 0	+ 1	+ 2

2	4	2	6	2	3	1	2	5	5
+ 6	+ 5	+ 7	+ 4	+ 8	+ 4	+ 8	+ 2	+ 5	+ 3

3	4	2	10	1	4	6	4	3	2
+ 7	+ 5	+ 3	+ 0	+ 3	+ 4	+ 3	+ 0	+ 6	+ 7

Name _____

Time Tests: Sums 0 - 10

Minutes

1	1/2	2	1/2	3	1/2	4	1/2	5	1/2
6	1/2	7	1/2	8	1/2	9	1/2	10	1/2

Color in your time!

```
  2     5     3     8     5     2     4     6     7     3
+ 0   + 5   + 2   + 2   + 3   + 3   + 4   + 3   + 2   + 3

  1     5     4     2     7     6     4     1     9     4
+ 1   + 2   + 1   + 4   + 3   + 2   + 2   + 6   + 1   + 3

  6     4     8     3     3     4     6     3     5     2
+ 0   + 4   + 1   + 7   + 4   + 3   + 1   + 3   + 4   + 3

  2     6     1     4     7     3     4     2     2     5
+ 5   + 3   + 7   + 2   + 3   + 1   + 3   + 1   + 2   + 3

  4     3     5     3     6     6     5     1     4     3
+ 6   + 5   + 1   + 6   + 1   + 4   + 2   + 2   + 5   + 0

  3     6     4     6     1     8     2     4     6     8
+ 7   + 3   + 4   + 2   + 3   + 2   + 7   + 5   + 3   + 2

  2     0     1     6     5     6     3     4     5     1
+ 8   + 9   + 4   + 4   + 5   + 3   + 5   + 4   + 5   + 6

  7     8     7     4     2     8     3     2     1     0
+ 0   + 2   + 3   + 4   + 7   + 2   + 6   + 1   + 5   + 7

  1     3     2     3     4     7     4     0     9     8
+ 9   + 4   + 4   + 2   + 4   + 2   + 6   + 6   + 1   + 2

  1     5     9     3     1     3     0     7     4     3
+ 1   + 4   + 1   + 3   + 8   + 4   + 9   + 2   + 4   + 2
```

©1991 Twin Sisters Productions, Inc.　　30　　TWIN 202 - Addition

2	7	4	8	3	5	8	6	2	4
+ 1	+ 3	+ 2	+ 4	+ 2	+ 3	+ 1	+ 2	+ 9	+ 6

3	7	2	6	7	3	5	2	8	3
+ 3	+ 5	+ 2	+ 1	+ 4	+ 7	+ 2	+ 8	+ 0	+ 9

7	4	8	3	4	2	5	9	8	2
+ 2	+ 3	+ 3	+ 1	+ 7	+ 3	+ 7	+ 1	+ 2	+ 7

4	6	2	5	3	6	4	6	5	3
+ 8	+ 5	+ 4	+ 4	+ 5	+ 3	+ 5	+ 6	+ 6	+ 6

2	4	9	5	7	4	6	3	2	9
+ 5	+ 4	+ 2	+ 5	+ 4	+ 8	+ 4	+ 4	+ 6	+ 3

Minutes

① ½ ② ½ ③ ½ ④ ½ ⑤ ½
⑥ ½ ⑦ ½ ⑧ ½ ⑨ ½ ⑩ ½

Color in your time!

5	3	5	7	6	4	5	4	6	3
+ 2	+ 9	+ 5	+ 3	+ 2	+ 5	+ 3	+ 2	+ 4	+ 6

5	9	3	5	6	4	7	5	7	8
+ 7	+ 1	+ 8	+ 2	+ 5	+ 4	+ 3	+ 4	+ 2	+ 4

3	5	1	3	7	9	9	4	3	7
+ 1	+ 6	+ 2	+ 5	+ 2	+ 2	+ 3	+ 3	+ 7	+ 4

6	7	4	8	6	4	2	8	7	8
+ 2	+ 5	+ 8	+ 1	+ 6	+ 6	+ 2	+ 4	+ 1	+ 2

4	8	3	9	3	8	6	4	3	9
+ 2	+ 3	+ 4	+ 2	+ 2	+ 2	+ 3	+ 7	+ 3	+ 3

Name _____

Minutes

① ½ ② ½ ③ ½ ④ ½ ⑤ ½
⑥ ½ ⑦ ½ ⑧ ½ ⑨ ½ ⑩ ½

Color in your time!

PROBLEMS CORRECT

2 + 6	7 + 5	3 + 8	6 + 6	3 + 1	6 + 2	4 + 7	9 + 1	2 + 4	5 + 3

5 + 7	4 + 5	2 + 3	7 + 4	9 + 3	3 + 7	8 + 2	4 + 1	6 + 5	5 + 2

2 + 8	8 + 4	6 + 5	3 + 2	2 + 9	6 + 3	10 + 2	2 + 2	7 + 2	3 + 6

3 + 5	4 + 4	5 + 5	4 + 2	8 + 3	3 + 4	4 + 8	9 + 2	5 + 4	4 + 6

2 + 7	5 + 6	3 + 3	6 + 4	4 + 3	2 + 1	5 + 3	3 + 9	10 + 1	2 + 5

3 + 3	7 + 2	8 + 4	3 + 5	2 + 1	6 + 4	10 + 2	4 + 2	5 + 6	0 + 6
6 + 5	4 + 1	2 + 2	8 + 2	5 + 4	4 + 2	6 + 3	4 + 8	2 + 3	3 + 9
4 + 3	3 + 6	2 + 4	6 + 2	3 + 8	0 + 1	2 + 3	8 + 4	1 + 8	10 + 2
6 + 6	8 + 4	7 + 3	2 + 4	3 + 2	6 + 1	4 + 4	6 + 2	3 + 7	0 + 5
7 + 1	3 + 3	4 + 5	3 + 6	5 + 7	6 + 6	2 + 5	7 + 5	8 + 3	9 + 1
9 + 1	0 + 2	2 + 6	3 + 3	3 + 5	0 + 8	5 + 6	4 + 6	3 + 4	7 + 2
2 + 7	4 + 3	3 + 4	9 + 3	7 + 3	4 + 7	6 + 6	2 + 1	5 + 5	6 + 2
7 + 4	4 + 4	4 + 8	9 + 1	2 + 8	8 + 3	2 + 1	3 + 3	0 + 3	5 + 4
5 + 2	0 + 4	2 + 9	10 + 2	9 + 1	2 + 2	3 + 2	4 + 6	6 + 6	7 + 5
5 + 5	5 + 3	4 + 1	7 + 0	3 + 1	3 + 3	8 + 2	0 + 9	7 + 3	6 + 2

Minutes

① ½ ② ½ ③ ½ ④ ½ ⑤ ½
⑥ ½ ⑦ ½ ⑧ ½ ⑨ ½ ⑩ ½

Color in your time!

WOW
PROBLEMS CORRECT

4 + 3	2 + 3	10 + 1	5 + 3	6 + 4	3 + 5	9 + 3	4 + 4	2 + 2	3 + 7
6 + 3	1 + 8	2 + 1	6 + 2	2 + 3	4 + 1	5 + 2	4 + 4	8 + 2	3 + 4
4 + 2	8 + 3	9 + 2	3 + 3	0 + 2	2 + 2	1 + 9	6 + 5	2 + 4	1 + 3
3 + 6	2 + 7	3 + 9	3 + 7	1 + 2	4 + 3	1 + 9	3 + 2	5 + 5	4 + 6
3 + 8	1 + 7	2 + 3	4 + 2	2 + 6	4 + 5	6 + 6	2 + 8	5 + 7	8 + 1
5 + 6	4 + 7	3 + 6	3 + 8	5 + 4	2 + 4	8 + 4	4 + 4	3 + 1	2 + 9
1 + 6	2 + 5	3 + 2	1 + 1	4 + 2	3 + 5	6 + 5	2 + 6	8 + 2	5 + 2
3 + 7	8 + 1	4 + 6	4 + 2	7 + 4	3 + 6	1 + 3	4 + 3	5 + 4	9 + 3
4 + 4	3 + 8	1 + 5	2 + 8	2 + 2	9 + 2	3 + 6	2 + 2	4 + 4	2 + 6
4 + 3	2 + 9	3 + 3	5 + 5	6 + 2	3 + 9	1 + 4	7 + 3	4 + 4	4 + 5

2 + 2	9 + 3	1 + 8	7 + 5	4 + 3	8 + 2	0 + 9	3 + 4	6 + 2	5 + 5
3 + 5	0 + 8	4 + 5	2 + 8	7 + 4	4 + 4	3 + 3	2 + 7	9 + 2	4 + 7
4 + 6	10 + 1	5 + 4	3 + 3	4 + 5	9 + 7	3 + 2	1 + 1	2 + 2	0 + 4
0 + 4	6 + 3	1 + 7	3 + 4	7 + 3	5 + 2	4 + 6	3 + 2	9 + 1	1 + 9
3 + 7	7 + 2	2 + 6	4 + 4	0 + 1	3 + 3	1 + 2	7 + 2	5 + 6	2 + 3
9 + 2	0 + 3	3 + 6	4 + 2	5 + 4	1 + 6	4 + 4	0 + 2	5 + 3	9 + 0
0 + 4	5 + 5	2 + 5	1 + 3	3 + 7	2 + 4	8 + 4	3 + 5	5 + 2	0 + 3
5 + 7	6 + 6	5 + 3	0 + 5	4 + 8	9 + 2	4 + 3	6 + 4	6 + 6	3 + 8
1 + 4	3 + 6	6 + 2	3 + 9	4 + 2	1 +7	8 + 3	5 + 3	0 + 6	5 + 6
6 + 6	10 + 1	0 + 7	6 + 3	3 + 8	2 + 9	6 + 5	4 + 5	1 + 5	8 + 2

36   TWIN 202 - Addition

6	3	5	2	8	7	4	6	2	6
+ 6	+ 1	+ 5	+ 2	+ 4	+ 7	+ 3	+ 4	+ 9	+ 5

7	8	4	9	6	2	7	9	3	9
+ 6	+ 5	+ 7	+ 6	+ 5	+ 3	+ 6	+ 4	+ 4	+ 5

3	7	2	8	4	8	4	8	9	2
+ 5	+ 9	+ 4	+ 9	+ 4	+ 3	+ 9	+ 8	+ 3	+ 8

3	1	9	4	7	2	6	3	8	4
+ 8	+ 5	+ 7	+ 5	+ 8	+ 5	+ 2	+ 6	+ 6	+ 6

4	2	9	6	3	8	4	9	2	3
+ 7	+ 7	+ 9	+ 2	+ 7	+ 7	+ 8	+ 8	+ 6	+ 9

Name _____

Minutes

① ½ ② ½ ③ ½ ④ ½ ⑤ ½
⑥ ½ ⑦ ½ ⑧ ½ ⑨ ½ ⑩ ½
Color in your time!

9	2	6	7	9	4	2	7	5	8
+ 9	+ 1	+ 3	+ 3	+ 1	+ 1	+ 5	+ 7	+ 2	+ 2

7	2	9	6	7	9	6	8	4	8
+ 4	+ 6	+ 2	+ 4	+ 2	+ 8	+ 5	+ 3	+ 5	+ 9

2	7	4	8	2	8	2	9	5	9
+ 2	+ 6	+ 2	+ 4	+ 4	+ 8	+ 7	+ 3	+ 3	+ 7

7	8	6	5	9	5	6	4	7	5
+ 4	+ 5	+ 3	+ 5	+ 4	+ 1	+ 6	+ 4	+ 6	+ 4

8	6	9	6	4	7	8	7	2	9
+ 6	+ 2	+ 5	+ 1	+ 3	+ 6	+ 7	+ 5	+ 3	+ 6

```
  9      6      3      7      2      9      5      8      3      9
+ 2    + 4    + 4    + 1    + 4    + 9    + 3    + 7    + 3    + 8
```

```
  2      5      9      1      7      3      8      3      6      6
+ 1    + 1    + 3    + 2    + 6    + 4    + 1    + 2    + 1    + 5
```

```
  8      3      8      5      7      2      8      5      8      7
+ 6    + 3    + 2    + 5    + 2    + 2    + 5    + 2    + 8    + 5
```

```
  6      9      6      7      2      6      9      2      4      8
+ 3    + 7    + 7    + 3    + 5    + 2    + 4    + 6    + 1    + 4
```

```
  8      4      7      9      5      8      6      7      4      9
+ 3    + 2    + 4    + 5    + 4    + 7    + 6    + 7    + 3    + 6
```

Name _____

Minutes

1 1/2 2 1/2 3 1/2 4 1/2 5 1/2

6 1/2 7 1/2 8 1/2 9 1/2 10 1/2

Color in your time!

1 + 8	6 + 7	9 + 1	2 + 5	7 + 2	3 + 4	5 + 1	2 + 3	8 + 1	3 + 6
2 + 8	4 + 1	0 + 4	5 + 9	9 + 2	1 + 5	2 + 1	6 + 6	2 + 2	10 + 6
7 + 3	12 + 3	9 + 3	2 + 2	3 + 3	5 + 2	6 + 2	8 + 2	1 + 4	2 + 7
6 + 8	0 + 5	3 + 5	12 + 4	6 + 5	8 + 2	4 + 2	10 + 2	3 + 2	7 + 4
5 + 7	5 + 3	9 + 4	2 + 6	4 + 3	12 + 2	7 + 5	6 + 4	2 + 4	5 + 8
11 + 3	7 + 6	2 + 2	4 + 6	9 + 5	2 + 9	11 + 4	4 + 4	0 + 2	8 + 4
5 + 6	2 + 6	9 + 6	11 + 5	0 + 3	6 + 3	2 + 5	8 + 5	5 + 4	3 + 1
9 + 7	2 + 3	7 + 5	5 + 5	8 + 6	12 + 5	11 + 6	7 + 7	0 + 1	11 + 7
7 + 4	5 + 5	8 + 7	4 + 4	4 + 5	6 + 2	9 + 8	2 + 4	10 + 5	7 + 3
2 + 7	6 + 6	0 + 6	2 + 1	9 + 9	11 + 7	4 + 4	4 + 8	2 + 8	8 + 8

40

WOW
___
PROBLEMS
CORRECT

Minutes
(1)(1/2) (2)(1/2) (3)(1/2) (4)(1/2) (5)(1/2)
(6)(1/2) (7)(1/2) (8)(1/2) (9)(1/2) (10)(1/2)
Color in your time!

8 + 9	2 + 1	7 + 3	6 + 9	1 + 4	8 + 3	3 + 2	12 + 5	9 + 2	4 + 5
4 + 6	9 + 3	5 + 5	2 + 2	6 + 8	7 + 1	6 + 2	8 + 4	5 + 6	11 + 3
12 + 3	8 + 8	6 + 3	9 + 5	0 + 4	9 + 1	2 + 3	10 + 8	1 + 5	7 + 4
10 + 3	2 + 4	7 + 2	3 + 1	4 + 4	7 + 6	9 + 4	11 + 5	6 + 7	3 + 3
1 + 6	9 + 7	5 + 6	0 + 4	8 + 1	6 + 4	2 + 5	1 + 1	7 + 5	4 + 3
10 + 4	8 + 2	6 + 6	3 + 6	0 + 3	4 + 9	8 + 8	1 + 7	10 + 2	3 + 4
8 + 7	2 + 7	10 + 4	6 + 6	3 + 5	1 + 4	5 + 7	4 + 2	0 + 1	7 + 9
5 + 8	7 + 8	10 + 3	1 + 8	9 + 9	10 + 5	2 + 8	6 + 5	3 + 6	5 + 2
2 + 2	0 + 8	3 + 7	6 + 2	2 + 9	8 + 5	7 + 7	5 + 9	9 + 6	4 + 8
8 + 6	3 + 8	7 + 7	10 + 5	4 + 7	4 + 1	3 + 9	0 + 4	1 + 9	7 + 8

Minutes

1 1/2 2 1/2 3 1/2 4 1/2 5 1/2
6 1/2 7 1/2 8 1/2 9 1/2 10 1/2

Color in your time!

5 + 2	7 + 1	4 + 3	1 + 2	11 + 4	7 + 9	2 + 2	6 + 3	7 + 5	3 + 2
8 + 8	9 + 7	5 + 3	7 + 2	3 + 3	8 + 6	5 + 8	6 + 9	2 + 3	7 + 4
4 + 4	0 + 5	1 + 3	8 + 7	2 + 4	6 + 2	10 + 4	9 + 1	5 + 7	4 + 1
9 + 6	2 + 5	8 + 2	12 + 5	7 + 8	1 + 4	9 + 8	6 + 6	3 + 4	4 + 9
3 + 5	5 + 4	1 + 5	7 + 3	6 + 4	9 + 5	10 + 3	8 + 5	2 + 6	6 + 8
2 + 1	4 + 4	6 + 5	6 + 6	2 + 7	0 + 4	5 + 5	1 + 6	9 + 9	3 + 6
5 + 6	0 + 8	1 + 7	4 + 6	4 + 4	5 + 9	8 + 8	0 + 3	4 + 8	11 + 5
2 + 8	8 + 3	7 + 6	3 + 8	2 + 1	9 + 4	1 + 8	8 + 9	9 + 9	3 + 7
3 + 9	0 + 7	6 + 6	1 + 9	3 + 3	12 + 4	1 + 6	4 + 5	2 + 9	8 + 4
9 + 3	4 + 4	0 + 6	3 + 2	7 + 5	2 + 2	9 + 8	6 + 7	7 + 7	1 + 1

Twin Sisters Productions
"Rap With The Facts" - Addition Rap

Side One with Answers - Side Two without Answers

If you have a number and you don't know what to do
When your teacher starts addition and you are so confused,
Start rappin' with this tape – don't fret – you'll understand,
'Cause we are gonna help you. We're a bad rappin' band.

Addition can be simple if you're adding 2 + 2,
But if you take it further, we'll tell you what to do.
Now just you keep on jammin'; we'll show you where it's at,
And, before you know it, you've memorized your facts.

Just to help you out we'll start with number 1,
And finish sums to 18 until we are done.
Are you ready to begin? Shout the answers out loud.
Snap your fingers, clap your hands, come on and join the crowd!

KIDS:
WE ARE ANXIOUS TO LEARN; WE WANT TO KNOW OUR FACTS.
WE'LL SHOUT THE ANSWERS OUT LOUD;
WE'LL HELP TO DO THE RAP.

The ones are first, you'll understand; listen to the song. Say the answers loud and clear; Smile and sing along.

1 + 1 = 2
1 + 2 = 3
1 + 3 = 4
1 + 4 = 5
1 + 5 = 6
1 + 6 = 7
1 + 7 = 8
1 + 8 = 9
1 + 9 = 10

Now you've got the rap; Are you ready to move on? The 2's are just the facts that we can build upon.

2 + 1 = 3
2 + 2 = 4
2 + 3 = 5
2 + 4 = 6
2 + 5 = 7
2 + 6 = 8
2 + 7 = 9
2 + 8 = 10
2 + 9 = 11

The 3's are kinda neat. Can you keep up with the beat? We're movin' very fast. Better hold on to your seat.

3 + 1 = 4
3 + 2 = 5
3 + 3 = 6
3 + 4 = 7
3 + 5 = 8
3 + 6 = 9
3 + 7 = 10
3 + 8 = 11
3 + 9 = 12

Are you ready for the 4's? Let's keep the beat alive. The music's getting funky; Just listen to it jive.

4 + 1 = 5
4 + 2 = 6
4 + 3 = 7
4 + 4 = 8
4 + 5 = 9
4 + 6 = 10
4 + 7 = 11
4 + 8 = 12
4 + 9 = 13

You know, as you move up the less we've got to learn. The 5's are next in line; They've been waiting for their turn.

5 + 1 = 6
5 + 2 = 7
5 + 3 = 8
5 + 4 = 9
5 + 5 = 10
5 + 6 = 11
5 + 7 = 12
5 + 8 = 13
5 + 9 = 14

The 6's I would like – to introduce to you. Let's buckle down and learn them well, and soon we'll be all through.

6 + 1 = 7
6 + 2 = 8
6 + 3 = 9
6 + 4 = 10
6 + 5 = 11
6 + 6 = 12
6 + 7 = 13
6 + 8 = 14
6 + 9 = 15

Listen to the music. Feel the beat in our song. Rap the 7's aloud. Now you're rappin' along.

7 + 1 = 8
7 + 2 = 9
7 + 3 = 10
7 + 4 = 11
7 + 5 = 12
7 + 6 = 13
7 + 7 = 14
7 + 8 = 15
7 + 9 = 16

Let's learn a few more facts. The problems are the 8's. I'm listening to your rap–ooh ... You sure are doing great!

8 + 1 = 9
8 + 2 = 10
8 + 3 = 11
8 + 4 = 12
8 + 5 = 13
8 + 6 = 14
8 + 7 = 15
8 + 8 = 16
8 + 9 = 17

The 9's, as you can see, have all been done before, So let's review the facts at hand and rap a little more.

9 + 1 = 10
9 + 2 = 11
9 + 3 = 12
9 + 4 = 13
9 + 5 = 14
9 + 6 = 15
9 + 7 = 16
9 + 8 = 17
9 + 9 = 18

And now that we have rapped the facts in order as they go,
A rockin' and a jammin' but perhaps a little slow,
Let's take the time and mix them up; we'll take it from the top.
Sums to 18 you'll have learned before we have to stop.

2 + 2 = 4
3 + 3 = 6
5 + 2 = 7
4 + 6 = 10
9 + 3 = 12
6 + 2 = 8

2 + 4 = 6
5 + 5 = 10
7 + 2 = 9
3 + 4 = 7
4 + 1 = 5
8 + 4 = 12

Keep Jumpin', Singin', Movin, Groovin'; Listen to it Rhyme.
Keep Dancin', Boppin', Rockin, Learnin', Rappin' all the Time.

If you practice faithfully, I'm sure that you will see,
Addition is as simple as learning your ABC's,
So, play your tape and dance along; memorize those facts.
I'm sure by now you realize that school is where it's at!

Keep Jumpin', Singin', Movin, Groovin'; Listen to it Rhyme.
Keep Dancin', Boppin', Rockin, Learnin', Rappin' all the Time.

 Twin 202 - Addition

MATH SKILLS AWARD

(3 + 5 =)

Presented To _____

For Scoring _____ *Within a* ____ *Minute*

Time limit on Addition Facts of _____ *(4 + 6 =)*

Date _____ *Signature* _____

GREAT EFFORT!

Presented To _____

For Scoring _____ *Within a* ____ *Minute*

Time limit on Addition Facts of _____

Date _____ *Signature* _____

ANSWER KEY

Page 3:

A.	5	**B.**	10	**C.**	7	**D.**	7	**E.**	10
F.	8	**G.**	4	**H.**	5	**I.**	9	**J.**	10
K.	8	**L.**	9	**M.**	10	**N.**	6	**O.**	8
P.	10	**Q.**	9	**R.**	10	**S.**	7	**T.**	10
U.	5								

Page 4:

A.	4	**B.**	5	**C.**	8	**D.**	7	**E.**	9
F.	9	**G.**	6	**H.**	6	**I.**	8	**J.**	10
K.	8	**L.**	10	**M.**	9	**N.**	8	**O.**	10

Page 5:

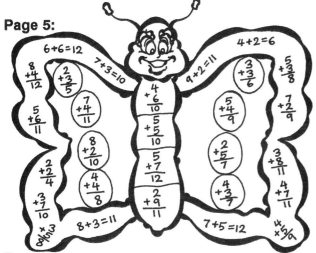

Page 6:

A. 7, 3, 11, 10, 2
B. 11, 5, 1, 9, 6
C. 8, 12, 4, 12, 10

Page 8:

A. 12, 6, 8, 8, 12, 11
B. 10, 7, 11, 11, 10, 4
C. 3, 9, 8, 12, 5, 7

Page 7:

A.	7	**B.**	12	**C.**	9	**D.**	11	**E.**	10
F.	9	**G.**	6	**H.**	4	**I.**	8	**J.**	5
K.	10	**L.**	11	**M.**	12	**N.**	8	**O.**	12
P.	10	**Q.**	8	**R.**	6	**S.**	10	**T.**	11

Page 10:

A.	4 + 2	**B.**	6 + 4	**C.**	4 + 1	**D.**	5 + 4	**E.**	6 + 6
F.	5 + 2	**G.**	4 + 0	**H.**	3 + 3	**I.**	4 + 4	**J.**	6 + 5

Page 12:

A.	12	**B.**	5	**C.**	8	**D.**	11	**E.**	15
F.	3	**G.**	2			**H.**	6	**I.**	16
J.	13	**K.**	7	**L.**	14	**M.**	10	**N.**	5

Page 9:

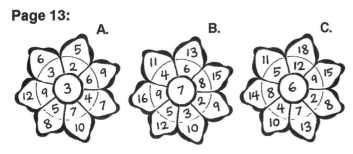

Page 11:

A. 12, 5, 3, 1, 14, 18
B. 6, 10, 7, 4, 16, 17
C. 9, 11, 2, 8, 13, 13

Page 14:

A. 12, 16, 13, 11, 12, 14, 13, 10, 6, 14
B. 18, 16, 11, 14, 15, 16, 9, 12, 10, 12

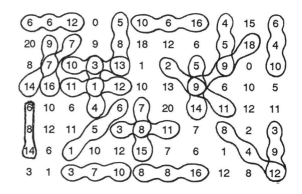

Page 13:

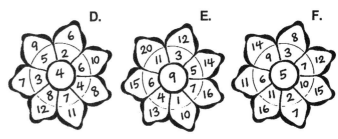

Twin 202 - Addition

ANSWER KEY

Page 15:

A. 11	B. 17	C. 16
D. 18	E. 17	F. 18
G. 16	H. 15	I. 18
J. 18	K. 10	L. 15

Page 16:

A. >	G. <	M. <	S. >
B. <	H. >	N. <	T. >
C. <	I. <	O. >	U. <
D. >	J. <	P. <	V. <
E. >	K. >	Q. <	W. <
F. <	L. <	R. >	X. <

Page 17:

A. 85	B. 91	C. 75	D. 74
E. 79	F. 89	G. 99	H. 68
I. 69	J. 86	K. 49	L. 90
M. 98	N. 66	O. 77	P. 87

Page 18:

A. 49	B. 79	C. 96	D. 37	E. 98
F. 97	G. 55	H. 85	I. 75	J. 69
K. 44	L. 89	M. 55	N. 57	O. 70
P. 73	Q. 91	R. 88	S. 26	T. 78

Page 19:

A. 87	B. 80	C. 60	D. 81	E. 91
F. 61	G. 65	H. 93	I. 92	J. 84
K. 64	L. 70	M. 90	N. 86	O. 62
P. 87	Q. 43	R. 95	S. 64	T. 86
U. 80	V. 85	W. 81	X. 61	Y. 90

Page 20:

59 +28		76 +19	41 +39		48 +28	65 +19	39 +56	29 +19	22 +29	38 +49	17 +34	26 +38
87		95	80		76	84	95	48	51	87	51	64
I		A	M		L	E	A	R	N	I	N	G

67 +28	19 + 9		16 +17	48 +17	28 +18	29 +23	36 +16	57 +19
95	28		33	65	46	52	52	76
A	T		S	C	H	O	O	L

Page 21:

A. 705	B. 410	C. 501
D. 722	E. 912	F. 802
G. 401	H. 916	I. 801
J. 920	K. 324	L. 433

Page 22:

A. 511	B. 981	C. 608	D. 522	E. 613
F. 420	G. 967	H. 831	I. 752	J. 523
K. 970	L. 510	M. 963	N. 995	O. 621
P. 793	Q. 321	R. 817	S. 506	T. 623

TWIN 202 - Addition

ANSWER KEY

Page 23:
- A. 5 + 2 = 7
- B. 4 + 4 = 8
- C. 3 + 2 = 5
- D. 9 + 1 = 10
- E. 4 + 2 = 6
- F. 7 + 2 = 9
- G. 2 + 2 = 4
- H. 1 + 2 = 3

Page 24:
- A. 6 + 4 = 10
- B. 9 + 3 = 12
- C. 4 + 2 = 6
- D. 8 + 2 = 10
- E. 3 + 1 = 4
- F. 5 + 7 = 12
- G. 11 + 6 = 17

Page 25:
- A. 2 + 4 = 6
- B. 6 + 5 = 11
- C. 3 + 7 = 10
- D. 8 + 4 = 12
- E. 2 + 7 = 9
- F. 1 + 2 = 3

Page 26:
- A. 2 + 3 = 5
- B. 6 + 5 = 11
- C. 7 + 2 = 9
- D. 9 + 8 = 17
- E. 4 + 6 = 10
- F. 11 + 5 = 16
- G. 2 + 5 = 7
- H. 12 + 6 = 18

Page 27:
- A. 12 + 5 = 17
- B. 4 + 6 = 10
- C. 9 + 7 = 16
- D. 3 + 2 = 5
- E. 8 + 10 = 18
- F. 6 + 5 = 11

Page 28:
- A. $3.25 + $5.50 = $8.75
- B. $5.95 + $4.39 = $10.34
- C. $5.99 + $5.95 = $11.94
- D. $4.99 + $3.25 = $8.24
- E. $4.99 + $4.39 = $9.38
- F. $5.99 + $5.50 = $11.49

Page 29:

6, 3, 10, 7, 10, 9, 7, 6, 7, 10

5, 6, 10, 5, 7, 6, 4, 10, 8, 7

7, 8, 10, 6, 8, 9, 7, 8, 4, 9

8, 9, 9, 10, 10, 7, 9, 4, 10, 8

10, 9, 5, 10, 4, 8, 9, 4, 9, 9

Page 30:

2, 10, 5, 10, 8, 5, 8, 9, 9, 6

2, 7, 5, 6, 10, 8, 6, 7, 10, 7

6, 8, 9, 10, 7, 7, 7, 6, 9, 5

7, 9, 8, 6, 10, 4, 7, 3, 4, 8

10, 8, 6, 9, 7, 10, 7, 3, 9, 3

10, 9, 8, 8, 4, 10, 9, 9, 9, 10

10, 9, 5, 10, 10, 9, 8, 8, 10, 7

7, 10, 10, 8, 9, 10, 9, 3, 6, 7

10, 7, 6, 5, 8, 9, 10, 6, 10, 10

2, 9, 10, 6, 9, 7, 9, 9, 8, 5

Page 31:

3, 10, 6, 12, 5, 8, 9, 8, 11, 10

6, 12, 4, 7, 11, 10, 7, 10, 8, 12

9, 7, 11, 4, 11, 5, 12, 10, 10, 9

12, 11, 6, 9, 8, 9, 9, 12, 11, 9

7, 8, 11, 10, 11, 12, 10, 7, 8, 12

Page 32:

7, 12, 10, 10, 8, 9, 8, 6, 10, 9

12, 10, 11, 7, 11, 8, 10, 9, 9, 12

4, 11, 3, 8, 9, 11, 12, 7, 10, 11

8, 12, 12, 9, 12, 10, 4, 12, 8, 10

6, 11, 7, 11, 5, 10, 9, 11, 6, 12

Page 33:

8, 12, 11, 12, 4, 8, 11, 10, 6, 8

12, 9, 5, 11, 12, 10, 10, 5, 11, 7

10, 12, 11, 5, 11, 9, 12, 4, 9, 9

8, 8, 10, 6, 11, 7, 12, 11, 9, 10

9, 11, 6, 10, 7, 3, 8, 12, 11, 7

Page 34:

6, 9, 12, 8, 3, 10, 12, 6, 11, 6

11, 5, 4, 10, 9, 6, 9, 12, 5, 12

7, 9, 6, 8, 11, 1, 5, 12, 9, 12

12, 12, 10, 6, 5, 7, 8, 8, 10, 5

8, 6, 9, 9, 12, 12, 7, 12, 11, 10

10, 2, 8, 6, 8, 8, 11, 10, 7, 9

9, 7, 7, 12, 10, 11, 12, 3, 10, 8

11, 8, 12, 10, 10, 11, 3, 6, 3, 9

7, 4, 11, 12, 10, 4, 5, 10, 12, 12

10, 8, 5, 7, 4, 6, 10, 9, 10, 8

TWIN 202 - Addition

ANSWER KEY

Page 35:

7, 5, 11, 8, 10, 8, 12, 8, 4, 10

9, 9, 3, 8, 5, 5, 7, 8, 10, 7

6, 11, 11, 6, 2, 4, 10, 11, 6, 4

9, 9, 12, 10, 3, 7, 10, 5, 10, 10

11, 8, 5, 6, 8, 9, 12, 10, 12, 9

11, 11, 9, 11, 9, 6, 12, 8, 4, 11

7, 7, 5, 2, 6, 8, 11, 8, 10, 7

10, 9, 10, 6, 11, 9, 4, 7, 9, 12

8, 11, 6, 10, 4, 11, 9, 4, 8, 8

7, 11, 6, 10, 8, 12, 5, 10, 8, 9

Page 36:

4, 12, 9, 12, 7, 10, 9, 7, 8, 10

8, 8, 9, 10, 11, 8, 6, 9, 11, 11

10, 11, 9, 6, 9, 16, 5, 2, 4, 4

4, 9, 8, 7, 10, 7, 10, 5, 10, 10

10, 9, 8, 8, 1, 6, 3, 9, 11, 5

11, 3, 9, 6, 9, 7, 8, 2, 8, 9

4, 10, 7, 4, 10, 6, 12, 8, 7, 3

12, 12, 8, 5, 12, 11, 7, 10, 12, 11

5, 9, 8, 12, 6, 8, 11, 8, 6, 11

12, 11, 7, 9, 11, 11, 11, 9, 6, 10

Page 37:

12, 4, 10, 4, 12, 14, 7, 10, 11, 11

13, 13, 11, 15, 11, 5, 13, 13, 7, 14

8, 16, 6, 17, 8, 11, 13, 16, 12, 10

11, 6, 16, 9, 15, 7, 8, 9, 14, 10

11, 9, 18, 8, 10, 15, 12, 17, 8, 12

Page 38:

18, 3, 9, 10, 10, 5, 7, 14, 7, 10

11, 8, 11, 10, 9, 17, 11, 11, 9, 17

4, 13, 6, 12, 6, 16, 9, 12, 8, 16

11, 13, 9, 10, 13, 6, 12, 8, 13, 9

14, 8, 14, 7, 7, 13, 15, 12, 5, 15

Page 39:

11, 10, 7, 8, 6, 18, 8, 15, 6, 17

3, 6, 12, 3, 13, 7, 9, 5, 7, 11

14, 6, 10, 10, 9, 4, 13, 7, 16, 12

9, 16, 13, 10, 7, 8, 13, 8, 5, 12

11, 6, 11, 14, 9, 15, 12, 14, 7, 15

Page 40:

9, 13, 10, 7, 9, 7, 6, 5, 9, 9

10, 5, 4, 14, 11, 6, 3, 12, 4, 16

10, 15, 12, 4, 6, 7, 8, 10, 5, 9

14, 5, 8, 16, 11, 10, 6, 12, 5, 11

12, 8, 13, 8, 7, 14, 12, 10, 6, 13

14, 13, 4, 10, 14, 11, 15, 8, 2, 12

11, 8, 15, 16, 3, 9, 7, 13, 9, 4

16, 5, 12, 10, 14, 17, 17, 14, 1, 18

11, 10, 15, 8, 9, 8, 17, 6, 15, 10

9, 12, 6, 3, 18, 18, 8, 12, 10, 16

Page 41:

17, 3, 10, 15, 5, 11, 5, 17, 11, 9

10, 12, 10, 4, 14, 8, 8, 12, 11, 14

15, 16, 9, 14, 4, 10, 5, 18, 6, 11

13, 6, 9, 4, 8, 13, 13, 16, 13, 6

7, 16, 11, 4, 9, 10, 7, 2, 12, 7

14, 10, 12, 9, 3, 13, 16, 8, 12, 7

15, 9, 14, 12, 8, 5, 12, 6, 1, 16

13, 15, 13, 9, 18, 15, 10, 11, 9, 7

4, 8, 10, 8, 11, 13, 14, 14, 15, 12

14, 11, 14, 15, 11, 5, 12, 4, 10, 15

Page 42:

7, 8, 7, 3, 15, 16, 4, 9, 12, 5

16, 16, 8, 9, 6, 14, 13, 15, 5, 11

8, 5, 4, 15, 6, 8, 14, 10, 12, 5

15, 7, 10, 17, 15, 5, 17, 12, 7, 13

8, 9, 6, 10, 10, 14, 13, 13, 8, 14

3, 8, 11, 12, 9, 4, 10, 7, 18, 9

11, 8, 8, 10, 8, 14, 16, 3, 12, 16

10, 11, 13, 11, 3, 13, 9, 17, 18, 10

12, 7, 12, 10, 6, 16, 7, 9, 11, 12

12, 8, 6, 5, 12, 4, 17, 13, 14, 2